AA002990

Microbes at Biomedical Interfaces 2019

Topical Conference at the 2019 AIChE Annual Meeting

Orlando, Florida, USA
10-15 November 2019

ISBN: 978-1-7138-0557-1

Printed from e-media with permission by:

Curran Associates, Inc.
57 Morehouse Lane
Red Hook, NY 12571

Some format issues inherent in the e-media version may also appear in this print version.

Printed with permission by Curran Associates, Inc. (2020)

For permission requests, please contact AIChE
at the address below.

AIChE
120 Wall Street, FL 23
New York, NY 10005-4020

Phone: (800) 242-4363
Fax: (203) 775-5177

www.aiche.org

Additional copies of this publication are available from:

Curran Associates, Inc.
57 Morehouse Lane
Red Hook, NY 12571 USA
Phone: 845-758-0400
Fax: 845-758-2633
Email: curran@proceedings.com
Web: www.proceedings.com

TABLE OF CONTENTS

(104B) INVITED SPEAKER: INNATE IMMUNITY AT THE BIOMATERIAL INTERFACE 1
Brian A. Pettygrove, Rachel M. Kratofil, Jovanka M. Voyich, Kyler B. Pallister, Paul Kubes, Philip S. Stewart

(104C) INVITED SPEAKER: STRUCTURE, STABILITY, AND MECHANICS OF BACTERIAL COMMUNITIES COLOCALIZED WITH FIBRIN NETWORK 2
Michael J. Solomon, Tianhui (Maria) Ma, J. Scott Van Epps

(104D) INVITED SPEAKER: MICROBIAL INTERACTIONS WITH THE INTESTINAL MUCOSAL BARRIER: IMPACT OF LUMEN STIMULI 3
Rebecca L. Carrier, Chia-Ming Wang, Jaclyn Lock, Taylor Carlson

(104E) INVITED SPEAKER: CONTROLLING BACTERIAL ADHESION ON BIOMEDICAL INTERFACES WITH BIO-INSPIRED LIQUID-INFUSED SURFACES 4
Caitlin Howell

(104F) INVITED SPEAKER: RESPONSIVE ANTIBACTERIAL AND ANTIFUNGAL BIOMATERIALS FOR INFECTION TREATMENT 5
Anita Shukla

(135A) LEVERAGING UV-C LEDS VS. MERCURY-BASED LAMPS FOR ELECTRONIC DISINFECTION IN HEALTHCARE 6
Norman Horn, Christian Davis

(135B) INVITED SPEAKER: THERMAL MITIGATION OF BIOFILMS 7
Eric Nuxoll

(135C) INVITED SPEAKER: ANTI-TOXIN STRATEGIES AS AN ANTIBIOTIC ALTERNATIVE 8
Evan Koufos, Eric Krueger, En Hyung Chang, Angela C. Brown

(135D) INVITED SPEAKER: PEPTIDE-BASED APPROACHES TO TREATING AND PREVENTING FUNGAL INFECTIONS 9
Amy J. Karlsson

(135E) INVITED SPEAKER: TOWARD COMPUTER-MADE ANTIBIOTICS 10
Cesar De La Fuente-Nunez

(135F) ACID-TRANSFORMING CHITOSAN (ATC)/DNA POLYPLEXES TO ELIMINATE INTRACELLULAR SALMONELLA TYPHIMURIUM INFECTION 11
Julius Edson, Weiping Chu, Steffen Porwollik, Michael McClelland, Young Jik Kwon

(135G) INVITED SPEAKER: ERADICATING ANTIBIOTIC-TOLERANT BACTERIA AND BIOFILMS BY ANTIMICROBIAL MODULAR PROTEINS PRODUCED BY CELL-FREE PROTEIN SYNTHESIS 12
Seok Hoon Hong

(135H) MECHANICAL VIBRATIONS AT RESONANCE FREQUENCY ON HYDROPHILIC SUBSTRATES ENHANCES BIOFILM REMOVAL 13
William Ballance, Inkyu Oh, Yang Lai, Mohamed Elhebeary, Taher Saif, Yuhang Hu, Hyunjoon Kong

(135I) BLOCK COPOLYMER NANOPARTICLES REMOVE BIOFILMS OF DRUG-RESISTANT GRAM-POSITIVE BACTERIA BY NANOSCALE BACTERIAL DEBRIDEMENT 14
Mary B. Chan-Park, Jianghua Li

(210B) HEAT EFFECTS ON VIABILITY, MORPHOLOGY, AND DISPERSAL OF STAPHYLOCOCCUS EPIDERMIDIS BIOFILMS 15
Joanne Beckwith, J. Scott Van Epps, Michael J. Solomon

(210C) THERMAL SHOCK DYNAMICS ON BIOFILM VIABILITY AT BIOMEDICAL INTERFACES 16
Haydar Aljaafari, Eric Nuxoll

(210D) STRUCTURING MICROBIAL BIOFILMS WITH 3D PRINTING 17
Reha Abbasi, James Wilking

(210E) THE ROLE OF BACTERIAL CHEMOTAXIS IN IMPROVING BIOREMEDIATION EFFICIENCY 18
Xueying Zhao, Roseanne M. Ford

(210F) ANTIBACTERIAL AND ANGIOGENIC MULTIFUNCTIONAL ADHESIVES FOR THE TREATMENT OF CHRONIC WOUNDS 19
Bahram Saleh, Thomas J. Webster

(210G) STUDY OF COLLAGEN TETHERED ANTIMICROBIAL PEPTIDE BINDING AND ANTIMICROBIAL ACTIVITY 21
Ziqi Wei, Lindsay D. Lozeau, Marsha W. Rolle, Terri A. Camesano

(210H) FUSOGENIC LIPOSOMES FUNCTIONALIZED WITH CELL PENETRATING PEPTIDES FOR THE TREATMENT OF BACTERIAL MENINGITIS 22
Caterina Bartomeu Garcia, Di Shi, Thomas Webster

(210I) ERADICATION OF BACTERIAL PERSISTER CELLS BY TARGETING MEMBRANE POTENTIAL 24
Sweta Roy, Ali Adem Bahar, Huan Gu, Dacheng Ren

(210K) MUCOID SWITCH ENABLES A GROWTH ADVANTAGE TO PSEUDOMONAS AERUGINOSAIN INTERFACIAL ENVIRONMENTS 25
Sricharani Balmuri, Nicholas Waters, Tagbo H. R Niepa

(210A) BIOFILM PREVENTION BY UVC SIDE EMITTING OPTICAL FIBERS 26
Mariana Lanzarini-Lopes, Sergi Garcia-Segura, Paul Westerhoff

(293A) INVITED SPEAKER: UNDERSTANDING VIRUS SURFACE INTERACTIONS AND STABILITY 28
Caryn L. Heldt, Xue Mi, Whitney C. Blocher McTigue, Mallory Bunker, Pratik U. Joshi, Sarah L. Perry

(293B) PREVENTING BACTERIAL BIOFILM FORMATION USING ACTIVE SURFACE TOPOGRAPHY 29
Huan Gu, Sang Lee, Dacheng Ren

(293C) INVITED SPEAKER: DEVELOPING TOOLS TO STUDY AND ENGINEER MICROBES BENEFICIAL TO HUMAN HEALTH 30
Nikhil U. Nair

(293G) INVITED SPEAKER: DESIGNING FUNCTIONAL NANOCULTURE SYSTEMS FOR STUDYING MICROBIAL DYNAMICS 31
Tagbo H. R Niepa, Huda Usman

(293E) MICROSPHERE HYDROGEL ENCAPSULATION TO IDENTIFY INTERACTIONS IN CHRONIC WOUND MICROBIAL CONSORTIA 32
Shawna Pratt, Petria Russell, Connie B. Chang

(293F) INVITED SPEAKER: DROP STABILIZATION ON A CHIP (DROPSOAC): STABILIZING MICROFLUIDIC DROPS FOR TIME-LAPSE QUANTIFICATION OF SINGLE-CELL BACTERIAL PHYSIOLOGY 33
Connie B. Chang

(293D) INVITED SPEAKER: DEVELOPING AND APPLYING A MICRODROPLET CO-CULTIVATION AND OMICS TOOLBOX FOR ELUCIDATING MICROBIOMES 34
Xiaoxia Nina Lin

(293H) INVITED SPEAKER: EFFORTS TOWARD MEASUREMENT ASSURANCE FOR MICROBIAL BIOFILM CHARACTERIZATION 35
Nancy J. Lin, Sandra M. Da Silva, Kirsten Parratt, Joy Dunkers

(316A) INVITED SPEAKER: MICROBIAL BIOFILMS: STRUCTURE, TRANSPORT, AND DYNAMICS 36
James Wilking

(316B) INVITED SPEAKER: ESTABLISHING AND EXPLOITING BIOCOLLOIDAL PROPERTIES OF BIOFILMS 37
Elizabeth J. Stewart

(316C) INOCULUM CONCENTRATION AS AN INDEPENDENT VARIABLE TO EVALUATE BIOFILM STRUCTURE-FUNCTION RELATIONSHIPS 38
Kirsten Parratt, Joy Dunkers, Heidi Leonard, Nancy J. Lin

(316D) INVITED SPEAKER: BIOFILM RHEOLOGY 39
Travis W. Walker

(316E) INVITED SPEAKER: MECHANICAL PRINCIPLES OF BIOFILM FORMATION 40
Jing Yan

(316F) INVITED SPEAKER: ROLE OF CHEMOTACTIC ADAPTATION IN H. PYLORI INVASION AND COLONIZATION 41
Pushkar Lele

(316G) DIATOM MICROBUBBLER FOR ACTIVE BIOFILM REMOVAL IN CONFINED SPACES 42
Yongbeom Seo, Jiayu Leong, Jun Dong Park, Yu-Tong Hong, Sang-Hyon Chu, Cheol Park, Dong Hyun Kim, Yu-Heng Deng, Vitaliy Dushnov, Joonghui Soh, Simon Rogers, Yi Yan Yang, Hyunjoon Kong

(428A) INVITED SPEAKER: THE BACTERIAL SENSE OF TOUCH: CHEMICAL AND MECHANICAL INTERACTIONS OF BACTERIAL CELLS AT INTERFACES 43
Maria M. Santore

(428B) INVITED SPEAKER: HOW BACTERIA READ THE MAP: EFFECTS OF SURFACE TOPOGRAPHY ON BIOFILM FORMATION 44
Dacheng Ren

(428C) INVITED SPEAKER: MOTILITY AND ADHESION OF BACTERIA NEAR OIL DROPLETS 45
Jacinta C. Conrad
(428D) INVITED SPEAKER: TRANSPORT PHENOMENA OF MICROBES IN GROUNDWATER AND POROUS MEDIA 46
Roseanne M. Ford
(428E) INVITED SPEAKER: GOODBYE HOSPITALS: HELLO IMPLANTABLE NANO SENSORS 47
Thomas J. Webster
(304A) CHEMICAL ENGINEERING INNOVATIONS FOR A RENEWABLE ECONOMY 48
Rakesh Agrawal
(304C) ENGINEERING MATERIALS FOR CLINICAL APPLICATION 49
Christine Schmidt
Author Index

570199 Invited Speaker: Innate Immunity at the Biomaterial Interface

This presentation will develop the hypothesis that early events in the innate immune response to a microbially contaminated implant determine whether a biofilm forms and a persistent infection occurs. A corollary of this hypothesis is that innate immunity can be locally guided or boosted to reduce infections on indwelling medical devices. To gain insight into the important phenomena in the interaction between a nascent biofilm and phagocytes, two systems were used to directly and dynamically image the activity of neutrophils against small aggregates of surface-attached *Staphylococcus aureus*. *In vitro*, human neutrophils were stained red and added to a serum-coated glass cover slip with pre-attached GFP-expressing *S. aureus*. *In vivo*, multiphoton confocal intravital imaging was accomplished in a mouse model in which a young *S. aureus* biofilm attached to a glass coverslip was interfaced to the bottom of the dermis in a mouse skin flap model. Bacteria were imaged via GFP, neutrophils by tdTomato (CatchupIVM-red mice), collagen by second harmonic generation, and vasculature by injected quantum dots. Neutrophil speed, migration, discovery of bacteria, and killing of bacteria were tracked and quantified through image analysis. Collectively these observations suggest that: 1) the timescale for arrival of neutrophils at the biofilm-affected material surface in this animal model is variable with a median longer than three hours, 2) bacteria continue to grow prior to neutrophil arrival both *in vitro* and *in vivo*, and 3) formation of bacterial aggregates makes it more difficult for neutrophils to clear the nascent biofilm.

573941 Invited Speaker: Structure, Stability, and Mechanics of Bacterial Communities Colocalized with Fibrin Network

Bacterial communities – biofilms – are a determinant of bacterial virulence in human infection. For example, biofilms that form on implanted medical devices are recalcitrant to treatment, including antibiotic regimens. Most experimental models of medical device infection have focused on biofilms that are monoculture; however, any such infection would be expected to be colocalized with host clot. Here we study the common blood borne pathogen and biofilm former *Staphylococcus epidermidis* in the presence of a host protein network comprised of fibrin, which is the primary structural component of clot. We report structure and mechanics on both microscopic and macroscopic scales. Bacteria present during fibrin network formation produce a visibly disorganized microstructure differing from the structure of a pure fibrin network. These structural changes corelate with fibrin network stiffness; they furthermore trigger mechanical instability over time. We also discuss the role of the staphylococcal exopolysaccharide, polysaccharide intercellular adhesin (PIA), on the stability of fibrin networks. Specifically, we use an *icaB* deletion mutant, which changes the charge state of PIA, to manipulate the degree of colocalization of fibrin and bacteria. These changes in colocalization affect mechanics and network structure. The results have potential implications for the persistence of surface adherent bacterial communities in the human circulatory system.

577854 Invited Speaker: Microbial Interactions with the Intestinal Mucosal Barrier: Impact of Lumen Stimuli

The intestinal mucosal barrier is highly significant to effective oral drug delivery, nutrient absorption, and interactions between microbes and underlying tissues. The mucosa is exposed each day to dynamic and variable intestinal lumen contents, yet the impact of these contents on the mucosal barrier is not well understood. Our laboratory is studying the impact of ingested materials, such as lipids in drug delivery systems or food, on transport through the intestinal mucosa of molecules (e.g., drugs and nutrients), particulates (e.g., drug carrier systems), and microbes. Results indicate that mild stimuli, such as those presented by food, can modulate the intestinal barrier, for example to impact oral drug delivery or microbial invasion, and that permeation through mucus is highly dependent on the physical and chemical properties of the penetrating material (drug, particle, microbe).

Particle and microbe tracking studies employing quantitative videomicroscopy revealed that mucus barrier properties are altered upon exposure to common food additive emulsifiers carboxymethylcellulose (CMC) and polysorbate 80 (Tween). Intestinal mucus was collected from porcine intestine and exposed to emulsifiers together with nanoparticles (passively diffusing entities) or *E. coli*as model mcirobes. Exposure to emulsifiers altered mucus structural and barrier properties. Overall, CMC and Tween significantly reduced effective diffusivity (Deff) of 200 nm particles with different surface chemistries. *E. coli*speed was decreased in mucus in the presence of CMC, but interestingly increased in the presence of Tween. *In vivo*exposure (injection into intestinal lumen) to CMC compacted the mucus layer, and Tween resulted in an increase in neutral and decrease in negatively-charged mucin sugars. Overall, results indicated that the emulsifiers significantly altered the local micro-environment within mucus, but it is not clear how these observed changes in static systems would translate into effects on a mucus layer on living tissue subjected to physiological exposure conditions (luminal flow in the presence of digestive enzymes and bile), or if an inflammatory response directly results from these acute exposure effects. We are working to clarify how these effects observed upon exposure of mucus to emulsifiers under static conditions translate into impact of transient exposure with luminal flow in the presence of representative luminal contents, which could potentially result in effective "clumping" and removal of mucus, and associated altered penetration of microbes, signaling molecules, and passively diffusing particles.

Exposure to dextran sodium sulfate (DSS), a chemical used to induce acute inflammation and epithelial damage driven by innate immune response to model IBD, results in denudation of the mucus layer. It has been demonstrated that bacteria are found within the "unpenetrable" inner mucus layer and adjacent to epithelium n this model prior to inflammation and indeed in all mouse models with spontaneous colitis and patients with active colitis. It is not clear if the denudation of the mucus layer results directly from acute exposure to DSS or is a downstream effect. In the present studies, porcine intestinal mucus was exposed to1% wt/vol DSS (20 ul on 200 ul mucus). Transport of passively diffusing nanoparticles was markedly decreased through mucus, yet average *E. coli*speed in mucus increased from 1.8 to 2.6 µm/sec, while *E. coli* speed in buffer was not affected by DSS. It is possible that the DSS is effectively "clumping" the mucus, possibly through cross-linking, thus creating larger channels through which microbes can swim. The concept of DSS "clumping" mucus, and potentially removing it from the epithelial surface, is supported by confocal images of cultures exposed to DSS.

These results indicate that mucus barrier properties are modulated by lumen stimuli, including stimuli presented by certain food substances.We hypothesize, given the crucial role of intestinal mucus in modulating interactions between intestinal contents and underlying tissues, that food substances directly impact the mucus barrier, and that an altered mucus barrier modifies interactions of microbes and signaling molecules with underlying tissues (e.g., proximity of microbes, binding of antimicrobial peptides).

One limitation to studying the intestinal mucosal barrier is lack of appropriate *in vitro* experimental models. Our laboratory is working to develop engineered intestinal models incorporating microbiota in homeostasis with intestinal epithelium and immune cells for studying the links between ingested material and intestinal homeostasis/inflammation.

569176 Invited Speaker: Controlling Bacterial Adhesion on Biomedical Interfaces with Bio-Inspired Liquid-Infused Surfaces

Bacterial interactions with surfaces are at the heart of many infection-related problems in healthcare. Recently, liquid-infused surfaces inspired by the *Nepenthes* pitcher plant have been introduced as a new approach to reducing surface adhesion. In this work, we describe the use of liquid-infused surfaces in the reduction of irreversible bacterial attachment both *in vitro* and *in vivo*. Tests on surfaces subjected to high concentrations of *Escherichia coli* and *Staphylococcus aureus* (10^6 cell/mL) for 48 h at 37° C showed reductions of up to 99% of attached biofilm and colony-forming unit counts 3 orders of magnitude lower on liquid-infused compared to control surfaces. Further investigation suggested that this was largely due to the spontaneous stripping away of a sacrificial layer of liquid at the surface when the sample was passed through an air/water interface. The reduction in bacterial cells was shown to persist in catheter-analog models where the liquid-infused surface was subject to bacterial cells under flow. Tests *in vivo* in murine model showed that hernia mesh implants which had liquid-infused surfaces showed fewer adherent bacteria than controls 3 days after inoculation with *S. aureus*, as well as a reduced immune response as indicated by capsule size and cytokine levels. The results presented here shed light on the interaction of bacteria with liquid interfaces, highlighting the fundamental differences between liquid-infused surfaces and traditional solid interfaces which may be exploited for future development of infection-resistant materials.

570121 Invited Speaker: Responsive Antibacterial and Antifungal Biomaterials for Infection Treatment

The World Health Organization has warned of an impending "post antibiotic era" in which common infections are lethal due to rising antimicrobial resistance. Microbes inherently evolve resistance mechanisms over time, the rates of which can be exacerbated by overuse of broad-spectrum antimicrobials and prolonged exposure at sub-inhibitory concentrations. Compounding this issue is the lack of discovery of new antimicrobial agents. I will describe our recent work on developing biomaterials, which can effectively treat microbial infections before further resistance can develop. Specifically, I will discuss the development of antimicrobial hydrogels and thin films that have been formulated to deliver small molecule antibiotics and antifungal agents, repurposed FDA-approved molecules with antimicrobial properties, and antimicrobial and antibiofilm peptides. We have formulated these highly tunable materials using modifications to synthetic and naturally-derived polymers, including gellan gum and poly(ethylene glycol). Our biomaterials exhibit a range of drug release behaviors (e.g., triggered-, swelling-, and diffusion-based release) and mechanical properties, and have demonstrated promising antimicrobial efficacy (eradication of gram-positive and gram-negative bacteria and several opportunistic fungi). I will focus primarily on describing (1) the microbial enzyme-triggered release of antifungal and antibiotic molecules from these materials and (2) the development of hydrogels exhibiting antibiofilm characteristics via the incorporation of antibiofilm peptides.

581358 Leveraging UV-C LEDs Vs. Mercury-Based Lamps for Electronic Disinfection in Healthcare

UV germicidal generators have been used commonly for both disinfection and treatment of water and surfaces for years. High-pressure or low-pressure mercury-based lamps are the most common devices used in these applications. However, mercury lamps come with many limitations including a long warm-up time, risk of mercury exposure, lower efficacy in refrigeration temperatures, heat generation, and broad light spectrum output. In order to minimize these limitations, UV-C LEDs operating at 260 nm to 280 nm were implemented to disinfect portable hand held electronics. Additional advantages of the restricted wavelength range of the UV-C LEDs include zero material degradation, minimal thermal transfer for battery preservation, thorough exposure in a shadow-less chamber, and minimal time (60 seconds) to desired efficacy LOG reduction.

574326 Invited Speaker: Thermal Mitigation of Biofilms

Destruction of biofilms under autoclave conditions (121 °C, ~1 hour) is well established but extraordinarily difficult to implement in vivo. Our laboratory investigates the thermal susceptibility of bacterial biofilms of common genera at sub-autoclave temperatures (50 – 80 °C) with various stationary phase population densities and architectures. Eventual elimination of the biofilm is correlated to a critical decrease in population density which is in turn correlated to the temperature and exposure time of the thermal shock. Further, antibiotic exposure works synergistically to eliminate biofilms at milder thermal shocks. Our laboratory also investigates approaches to delivery of these thermal shocks, such as magnetic nanoparticle / polymer coatings which can deliver these thermal shocks wirelessly from outside the body by applying an external alternating magnetic field. Design parameters for these coatings, as well as thermal modeling to minimize damage to adjacent tissue, will also be discussed.

569151 Invited Speaker: Anti-Toxin Strategies As an Antibiotic Alternative

Pathogenic bacteria produce numerous virulence factors, including protein toxins, to enhance their growth and survival within the host. Our lab seeks to understand the mechanisms of bacterial toxin delivery to identify and exploit therapeutic targets. In recent years, we have focused on a leukotoxin (LtxA) secreted by *A. actinomycetemcomitans,* which specifically kills human white blood cells to disrupt the host immune response and therefore plays a key role in bacterial colonization of the host. Disruption of the mechanism of action of this toxin would eliminate this immunosuppressive function, thus rendering the bacteria susceptible to natural clearance mechanisms. The cell type specificity of LtxA arises from its reported targeting of the lymphocyte function-associated antigen-1 (LFA-1) integrin, which is only expressed by human white blood cells. In addition, we recently showed that LtxA binds with a strong affinity to cholesterol, and inhibition of this binding to cholesterol prevents LtxA-mediated cytotoxicity. Using our knowledge of these essential membrane interactions, we have designed several targeted inhibitors of LtxA binding to host cells. In addition, we have uncovered the mechanism of toxin inhibition by catechins, a class of polyphenolic compounds found in tea. Together, these projects demonstrate the potential of using mechanistic information about toxin membrane interactions to design anti-toxin strategies to treat bacterial associated diseases.

579793 Invited Speaker: Peptide-Based Approaches to Treating and Preventing Fungal Infections

*Candida albicans*is a commensal organism of humans that is also a frequent cause of disease in immunocompromised patients. Infections caused by *C. albicans*include oral infections like oral thrush and biofilm-associated infections on medical devices. Treating infections caused by *C. albicans*can be challenging due to increasing drug resistance and the limited number of antifungal agents that are currently available. To address the current challenges associated with treating fungal infections, we are designing peptides capable of specifically targeting *C. albicans*. One of these peptides is histatin 5, an antimicrobial peptide found in human saliva that has potent and specific activity against *C. albicans*. Although histatin 5 has promise as a therapeutic agent, the fungus produces secreted aspartic proteases that degrade the peptide and reduce its antifungal activity. We have designed analogs of histatin 5 that offer strong resistance to the fungal proteases and improved antifungal activity. We are investigating the use of these analogs for treating and preventing *Candida* infections, including biofilm-associated infections, while also improving understanding of how the secreted aspartic proteases recognize and cleave peptide substrates. In addition to exploring peptides as potential therapeutics, we are also using peptides to target and deliver bioactive molecules to *Candida*pathogens. We are examining how properties of these cell-penetrating peptides affect their translocation across the cell wall and cell membrane of *Candida*cells and are defining the types of cargo they can carry into fungal cells. Our peptide-based strategies offer a new approach to designing antifungal therapeutics capable of treating and preventing fungal infections.

580273 Invited Speaker: Toward Computer-Made Antibiotics

Machines hold the potential to replace humans in many societal endeavors, and drug discovery is no exception. Antibiotic innovation has been stalled for decades, which has coincided with an alarming increase in multidrug-resistant bacteria. Since the beginning of the antibiotic era, the natural world has been our greatest innovator, giving rise to nearly all antibiotics available today. As mere observers of the vast molecular diversity produced by Earth's organisms, we have perfected the art of isolating novel chemistries with life-saving antimicrobial properties. However, today we are at a crossroads as no new molecular scaffolds have been discovered for decades. We may need to look beyond the natural world into the virtual dimension for solutions and harness present-day computational power to help solve the global health grand challenge of antibiotic resistance. Computer-made drugs may enable the discovery of unprecedented functions in biological systems and help replenish our arsenal of effective antibiotics. Proteins perform the cellular tasks required for life. The great variety of their biological activity is due in part to their vast combinatorial space: 20^n, n being the number of amino acids present in any given peptide chain and 20 being the number of natural amino acid monomer building blocks. Yet we do not have the tools to properly engineer these diverse molecules. One approach is to start small: I will present foundational frameworks to computationally develop peptides, tiny proteins that display great sequence diversity but are more amenable than larger molecules to redesign and engineering. My approach is to expand nature's repertoire to build novel synthetic peptides with extremely useful properties.

576317 Acid-Transforming Chitosan (ATC)/DNA Polyplexes to Eliminate Intracellular *Salmonella Typhimurium* infection

Drug-resistant infectious bacteria are a pressing societal challenge with conventional antibiotics are becoming less effective at treating bacterial infections. More specifically, gram-negative intracellular infection provides an additional challenge. An innovative therapy for combating these emerging superbugs requires a paradigm-shift in therapy development. Therefore, we engineered a combinatory therapy of nanoantibiotics and gene therapy. The need for an antimicrobial material with gene delivery capabilities led to the use of chitosan as our selected polymer. Here we use acid-transforming chitosan (ATC), a derivative of chitosan with enhanced solubility and pH-responsive capabilities. We explored the antimicrobial capabilities of ATC and determined the possible mechanism for the antimicrobial efficacy of ATC and chitosan. The efficacy of the free polymer was observed against S. typhimurium, an intracellular pathogen, alone at varying pH conditions which lead to a positive correlation between pH and effects the polymer had on the microbe. Additionally, ATC/DNA polyplexes were applied to RAW 264.7 cells infected with S. typhimurium expressing GFP and resulted in a decrease in the levels of bacteria found within the cells as confirmed by flow cytometry. The study indicated the possibility of using an antimicrobial vector against difficult intracellular infections and the advantages of the multi-faceted therapeutic.

575532 Invited Speaker: Eradicating Antibiotic-Tolerant Bacteria and Biofilms By Antimicrobial Modular Proteins Produced By Cell-Free Protein Synthesis

Chronic bacterial infections are difficult to treat through standard antibiotics because a small population of the bacteria in biofilms becomes tolerant to antibiotics by entering a dormant state. Such antibiotic tolerant bacteria are known as persisters and pose a significant problem in clinical settings. As antibiotic levels drop post-treatment, persister cells revert back into actively growing cells resulting in recalcitrant chronic infections. Currently, there are very limited approaches available for eradicating persister cells. Colicins, a type of antimicrobial bacteriocins, are considered as a viable alternative of conventional antibiotics due to their unique cell killing mechanisms that can damage cells by pore-forming on the cell membrane, nuclease activity, and cell wall synthesis inhibition. Additionally, colicins' cell killing activity is target-specific by recognizing receptors of the target cells without influencing other bacteria. Thus, colicins are excellent candidates for killing persisters. We utilized cell-free protein synthesis (CFPS) to produce colicins, as CFPS offers rapid protein production and characterization of such toxic proteins without cell-viability constraints and the need for protein purification. We demonstrated that colicins (*e.g.*, E1 and E2) produced by CFPS based on crude *Escherichia coli* lysates are active in eradicating persister cells by up to six orders of magnitude of cell viability reduction. Optimizing CFPS reactions such as including molecular chaperones and immunity proteins enabled significant improvement of solubility and activity of colicins. Colicin treatment was also effective in killing biofilm cells. Colicin production in the CFPS setting and cell-killing kinetics of cell-free synthesized colicins were very rapid, which provides a great potential for point-of-care drug producing systems by combining with lyophilized or paper-based CFPS. We further discuss the modular design of colicins that can redirect activity of colicins to target other bacteria and developing beneficial microbes that can eradicate or kill harmful biofilm cells via controlled colicin secretion. This study finds that colicins can eradicate antibiotic-tolerant bacteria and biofilm cells and that CFPS is a promising platform for rapid production, optimization, and characterization of toxic proteins.

574352 Mechanical Vibrations at Resonance Frequency on Hydrophilic Substrates Enhances Biofilm Removal

Biological fouling caused by biofilms poses a grave concern for both human health and industrial infrastructure by causing infection, corrosion of metals, failure of medical and industrial equipment, and persistent drag on marine ships. External mechanical vibrations are commonly used to clean bio-fouled substrates. However, vibration conditions are often chosen without an in-depth understanding of structure-biofilm interactions. As such, this study interrogates the integrative effects of the surface energy of a substrate and vibration conditions including frequency and voltage-induced energy input mode on removing biofilms.

For this study, we assembled a polypropylene substrate coated with a polyacrylamide hydrogel that does not weaken during vibration and attached it to a dielectric elastomer that causes controlled, continuous deflection. The polyacrylamide gel was engineered to have a high surface energy of around 73.5 mN m^{-1} and present a low number of hydrophobic crosslinks (i.e., 1:10800 molar crosslinker ratio). Both of which contribute to minimizing bacterial adhesion by reducing the number hydrophobic regions that cell-adhesive biomolecules can deposit. The tough gel coating was effective to lower the adhesion force of biofilms from 666 to 5.8 Pa (before and after coating) as measured by atomic force microscopy (AFM). Furthermore, the coated substrate could self-remove mature biofilms cultured up to 7 days when vibrated at the resonance frequency of the substrate. Using this set-up, we discovered that vibration at the structural resonance frequency using a square wave-form resulted in the highest deflection of the surface and maximizes the removal of biofilms. The findings from this study can be broadly applicable to the enhancement of various self-cleaning medical and industrial devices.

555517 Block Copolymer Nanoparticles Remove Biofilms of Drug-Resistant Gram-Positive Bacteria By Nanoscale Bacterial Debridement

Antimicrobial resistance has become a global healthcare crisis. Compounded with the evolution of multi-drug resistance, bacteria also develop biofilms to protect themselves, so that biofilm-associated infections are extremely difficult to treat. Antibiotics can be up to x1000 less effective to biofilms as compared with the planktonic form. Once developed into the biofilm form, it will become highly resistant to conventional antibiotics. Many antibiotics, natural antimicrobial peptides and synthetic antimicrobial agents have been studied as anti-biofilm agents but the general efficacy is still not high. Further, they usually suffer from the problem of toxicity and limited life span. Recently, we have developed a novel antibiofilm cationic copolymer nanoparticle that has excellent biofilm removal capability; the nanoparticle is self-assembled from dextran-*block*-(poly((3-acrylamidopropyl) trimethylammonium chloride(AMPTMA)-co-butyl methacrylate (BMA)) (DA95B5). Interestingly, the polysaccharide-based amphiphilic copolymer DA95B5 self-assembles into nanoparticle form which does not have any antibacterial effect but exhibits excellent preformed biofilm removal ability. As the cationic charge and hydrophobic segments are hidden inside the core of nanoparticles, the latter does not have any bacterial killing ability; however, the antifouling shell of the polysaccharide as well as the nanoparticles form, enhances the biofilm dispersal ability by a mechanism we termed "nanoscale bacterial debridement". Cryo-TEM and confocal microscopy show that the nanoparticle can penetrate into the biofilm and form a coating around the negatively charged bacteria to weaken the cell-biofilm interaction. Our *in vitro* results showed that the polymeric nanoparticles exhibit antibiofilm ability towards several multi-drug resistant and clinically relevant Gram-positive bacterial strains, with efficacy much higher and/or similar to the conventional standard antibiotics. Our in vivo data corroborates that such nanoparticles possess MRSA biofilm removal efficacy that is higher than vancomycin. Further, both *in vitro* and *in vivo* data showed our nanoparticles have good biocompatibility with low hemolysis and cytotoxicity. Overall, this novel biofilm removal platform of a cationic non-fouling nanoparticle provides exciting opportunities for treatment of multi-drug resistant biofilm infections which may have widespread applications.

577439 Heat Effects on Viability, Morphology, and Dispersal of *Staphylococcus Epidermidis* Biofilms

Bacterial biofilms are a leading cause of hospital-acquired infections and pose a unique challenge to treat. One alternative to the microbiological view of biofilms is to investigate them from the perspective of soft matter. In the context of soft matter, biofilms are analogous to a composition of colloids embedded in a cross-linked polymer gel. The development of biofilms on medical devices can be understood as a consequence of adsorption, growth, and detachment; each effect is governed by self-assembly, fluid mechanics, and transport phenomena which are dependent on fundamental chemical engineering principles. Here we show how fluid dynamics and heat transfer can be applied to prevent and/or remove bacterial biofilms in an *in vitro*, flow cell model of a dialysis catheter. Using *Staphylococcus epidermidis*, the most common bacterial species isolated from infected medical devices, we grow biofilms under physiologically relevant flow conditions and expose them to thermal degradation using a pre-warmed fluid that is injected into the model catheter. Using confocal microscopy and image analysis techniques, we establish that exposing biofilms to elevated temperatures causes a significant decrease in the viability of surface-adhered bacteria, as well as an increase in the structural heterogeneity. Additionally, the cellular material debrided during heat exposure was collected and assessed for morphology and viability. Finally, we investigated the coupled effects of heat treatment with the antibiotic vancomycin, on both the surface-adhered biofilm as well as the downstream debrided material. Understanding the response of both surface-adhered and dispersed bacterial cells under thermal stress, is a promising step toward the development of an *in situ* treatment/remediation method for biofilm growth in medical devices and provides new insight on the infection capabilities of disseminated biofilm material.

572615 Thermal Shock Dynamics on Biofilm Viability at Biomedical Interfaces

When bacteria colonize a medical implant surface, they form a biofilm which cannot be eradicated chemically. The current standard of care is surgical explantation of the device and surrounding tissue, with eventual reimplantation of a replacement device with twice the probability of infection. These infections are a $5 billion problem in the U.S. alone, impacting over 100,000 patients annually. One approach to mitigating these biofilms is *in situ* thermal eradication by generating a thermal shock directly at the device/biofilm interface. We have developed magnetic nanoparticle / polymer composite coatings which heat rapidly when exposed to an externally applied alternating magnetic field, and have quantified immediate bacterial population reductions of up to 6 orders of magnitude (i.e., "6 log" reductions).

Moreover, under certain circumstances the biofilm bacterial population continues to decrease for hours after the thermal shock is removed and physiological temperature is restored, resulting in complete eradication of the biofilm. Re-incubation studies at a variety of shock temperatures (50 – 80 °C) and exposure times (1 – 30 min) showed that thermal shocks producing immediate population reductions greater than a critical drop resulted in complete elimination of the biofilm several hours later.

Furthermore, this approach may be integrated with other modifications to the biofilm/device interface to prevent or eliminate biofilms with even milder thermal shocks. Investigations of combined thermal shock and antibiotic exposure yielded orders of magnitude population reductions greater than the sum of the orders of magnitude reduction for either treatment alone, enabling significantly milder thermal shocks to provide the same efficacy and providing an impetus for localized delivery at the device interface of antibiotics which are by themselves insufficient for biofilm elimination.

These investigations were done on biofilms of two different types of bacteria (*Pseudomonas aeruginosa* and *Staphylococcus aureus*), which were both cultured using two different protocols (shaker table and drip flow reactor). Despite the type of bacteria or the substantially larger difference in the initial population density due to the culturing protocol, an immediate critical population drop was required to prompt eventual die-off of the biofilm. Using antibiotics make this critical drop achievable at milder thermal shock. This suggests that biofilms on medical implant devices can be thermally mitigated at milder thermal shock than previously believed.

573393 Structuring Microbial Biofilms with 3D Printing

Most micro-organisms, including bacteria, fungi, and archaea, exist in surface-adherent, multicellular communities known as "biofilms". These films exhibit enhanced antibiotic resistance and reduced clearance by the immune system and are implicated in a wide variety of medical issues, including hospital-acquired and biomedical implant infections. Biofilm formation and virulence is largely governed by cell-to-cell communication, which is strongly influenced by community structure. However, structure-communication relationships remain poorly understood, and there is a need for technologies that provide well-defined spatial control over microbial structures. Here, we will present a light-based 3D printing platform for structuring bacteria in a biocompatible hydrogel. To explore the impact of spatial location on complex microbial interactions in biofilms, we print biomedically-relevant bacteria like *Pseudomonas aeruginosa* and follow their time-dependent behavior using confocal microscopy. We find that microorganisms in the 3D printed structures exhibit complex and unexpected dynamics. Knowledge gained from these experiments can be used to optimize structure-function relationships and may inform new strategies for biofilm disruption.

577261 The Role of Bacterial Chemotaxis in Improving Bioremediation Efficiency

Crude oil components exposure can cause potential health effect to oil spill cleanup workers. An alternative strategy to clean up oil spill is bioremediation. Bioremediation is a technique to use microorganisms to break down environmental pollutants. Microorganisms can accumulate near the oil-water interface, forming biofilms. This accumulation increases the bioavailability of dispersed oil droplets to hydrocarbon-degrading marine bacteria. For low solubility hydrocarbons, such as hexadecane, the formation of biofilm in an oil-water interface is particularly important. A process that contributes to biofilm formation is chemotaxis. In this process, marine bacteria sense the hydrocarbon concentration gradient in the oil-water interface and swim toward the higher concentration gradient. Chemotaxis can compensate for the mass-transfer limitation of the low solubility hydrocarbons, thereby enhancing biodegradation efficiency.

Marine bacteria chemotaxis is not well characterized, therefore, our work is to quantify how much chemotaxis contribute to increasing bioavailability of oil droplets to hydrocarbon-degrading bacteria. To analyze chemotaxis, we exposed marine bacteria to hexadecane in a uniquely designed microfluidic device that ensures the migration of bacteria is solely in response to a constant gradient of hexadecane. This device allows us to readily quantify chemotaxis from the resulting bacterial distribution. Experimental data was then used to quantify parameters in a mathematical model capturing motility and chemotaxis derived from first principles. Specifically, the random motility coefficient was obtained from experiments in a hydrocarbon-free condition and chemotaxis parameters were determined from experiments at different hexadecane concentrations. We found the random motility coefficient for marine bacteria in the model is in the same order of magnitude as that of *Escherichia coli*, suggesting the model is reasonable. This research enables us to understand the extent to which bacterial chemotactic processes facilitate recycling of dissolved hydrocarbons. Further characterization over different marine bacteria can identify potential bioreporters to sense oil spill in seawater, hence, decreasing human labor and reducing human exposure to crude oil.

575888 Antibacterial and Angiogenic Multifunctional Adhesives for the Treatment of Chronic Wounds

Antibacterial and Angiogenic Multifunctional Adhesives for the Treatment of Chronic Wounds

Bahram Saleh[1,2], Thomas J. Webster[1,2*]

[1]Department of Chemical Engineering, Northeastern University, Boston, MA 02115

[2]Nanomedicine Science and Technology Center, Northeastern University, Boston, MA 02115

Statement of Purpose: Diabetes mellitus is an important health problem that affects millions of people worldwide. Diabetic patients suffer from skin chronic wounds, which do not heal in a timely and orderly fashion. These wounds are characterized by impaired vascularization, uncontrolled inflammation, and bacterial infection. In particular, excessive inflammation impairs the healing process, reduces the protective role of immune cells, and could lead to bacterial infection. On the other hand, bacterial colonization exacerbates prolonged inflammation and causes tissue injury due to the interaction of the host with multiplying bacteria. Moreover, the decreased blood supply to full thickness wounds impairs the immune response by reducing the infiltration of leukocytes and macrophages at early stages of inflammation. Therefore, there is an unmet clinical need to develop innovative therapeutic strategies to overcome these overlaping challenges in chronic wounds. We designed a biocompatible, biodegradable, and photocrosslinkable adhesive hydrogels by introducing mussel adhesive moiety (catechol) and methacrylic anhydride (MA) to a naturally derived polymer, gelatin. Also, we developed an antimicrobial-hybrid Zinc-Oxide@Gold (ZnO@Au) tetrapods. By dispersion of these tetrapods in the 3D networks of the hydrogels we attain nanocomposite hydrogels that can be applied readily on the wound site and be crosslinked in a few seconds. Moreover, ZnO@Au tetrapods can enhance angiogenesis in the wound area according to proangiogenic properties of ZnO.

Methods: Antimicrobial and angiogenic tetrapods were synthesized using a wet chemical route and coated with PEG-DA (dopamine conjugated to 500 kDa polyethyleneglycol) to conclude water soluble tetrapods. Catechol modified photocrosslinkable pre-polymer for the hydrogel formation was synthesized in two consecutive reactions. The crosslinking reagents BOP ((Benzotriazol-1-yloxy) tris(dimethylamino)phosphonium hexafluorophosphate) and HOBt (Hydroxybenzotriazole) were used to conjugate catechol groups to the gelatin backbone and then catechol modified gelatin was precipitated and washed in excess acetone. Next, the catechol modified gelatin was dissolved in carbonate/bicarbonate buffer (pH=9.5) and methacrylated in a direct reaction with methacrylic anhydride. To prevent the undesired crosslinking during the reaction, we performed the reactions under an inert atmosphere of nitrogen. Different concentrations of tetrapods were mixed with hydrogel precursors in PBS (phosphate buffer saline), containing triethanolamine and N-vinylcaprolactam. Eosin Y disodium salt was dissolved separately in phosphate buffer saline and used as the photoinitiator. The tetrapod/hydrogel precursor solution was then mixed with Eosin Y and photopolymerized. The adhesive properties of hydrogels were assessed using standardized tests (e.g., ASTM F2392-04), and compared to commercially available tissue adhesives such as Evicel and Coseal, Colony counting test was performed using different concentration of tetrapods and tetrapods in hydrogels to assess the antimicrobial properties of them. Human dermal fibroblast (HDF) cells were used as model cells to measure the cytotoxicity of the tetrapods. Next, hydrogels were formed on the membrane of transwell inserts, and the inserts were placed into the wells containing the human umbilical vein cells (HUVECs). The intensity of CD31 expression on HUVECs was quantified from fluorescent images of anti-CD31 stained cells.

Results: We first synthesized tetrapod loaded adhesive hydrogels using varying concentrations of tetrapods. We optimized the mechanical and physical properties of the engineered adhesives for skin tissue regeneration (elastic modulus of around 130 kPa, extensibility more than 100). Our results showed strong adhesion of tetrapod loaded hydrogels in a wound closure test using pig skin as substrate. The adhesion strength of the tetrapod/hydrogels (148.4 ± 14.5 kPa) was significantly higher than those commercially available adhesives, Evicel® (i.e., 26.3 ± 4.7 kPa) and Coseal® (19.4 ± 17.3 kPa). Tetrapods showed antibacterial activity against different multidrug resistant bacteria strains in the concentration as low as 10 µg/ml. Quantification of the intensity of fluorescently labeled CD31 on HUVECs demonstrated around 1400 times increase in the expression of CD31, as compared to control HUVECs incubated in culture media, at day 1 and 3 of treatment with tetrapods.

Conclusions: Our results demonstrated that the engineered hydrogels can be potentially used as a multifunctional adhesive, with the ability to control infection, promote angiogenesis, and consequently tissue healing. In addition, due to its strong adhesion to skin and rapid closure of the wound, engineered hydrogels can cover skin wounds in any shape and size, especially in cases that no suture can be used.

565165 Study of Collagen Tethered Antimicrobial Peptide Binding and Antimicrobial Activity

Recent work from our lab has demonstrated that collagen tethering of antimicrobial peptides (AMPs) using collagen binding domains (CBDs) has potential in delivering novel, non-cytotoxic, antimicrobial, and pro-healing therapies to chronic wounds and will not promote bacterial resistance. Our lab designed two chimeric versions of human-derived LL37 with C-terminal CBDs derived from collagenase (*c*CBD-LL37) and fibronectin (*f*CBD-LL37) and we found that the peptides were antimicrobial but not cytotoxic when adsorbed onto collagen. However, optimizing the formation of the deposited collagen films, in terms of the appropriate number of binding sites and optimal loading CBD-LL37 concentration for treating different types of bacteria needs to be further studied. The goal of this project is to study collagen film deposition, to quantify CBD-LL37 binding concentration, retention over time and adsorbed surface density onto collagen type I, and to observe antimicrobial activity on *E. coli* and Methicillin-resistant *S. aureus* in quartz crystal microbalance with dissipation monitoring (QCM-D). The frequency and dissipation are two main measurements of QCM-D which can reflect the mass adsorption and the viscoelastic properties of the film, respectively. To realize our goal, we first studied the concentration-dependent viscoelastic deposition of collagen type I films onto SiO_2 or Au substrates by measuring frequency and dissipation in QCM-D. We then studied time- and concentration-dependent interactions of CBD-LL37 with these films and quantified the theoretical number of binding sites between CBD-LL37 and collagen. To study the antimicrobial activity, we introduced *E. coli* or Methicillin-resistant *S. aureus* into QCM-D over AMP functionalized collagen and monitored real-time changes in frequency and dissipation. We also performed *ex situ* antimicrobial assays which consisted of sample preparation, plate culture, and colony counting steps to determine the antimicrobial activity of collagen tethered CBD-LL37. Collagen films were then visualized *ex situ* using SEM. The results of these studies will help to optimize concentration-dependent tethering of CBD-LL37 onto collagen-based substrates and provide insight on the best approach to using collagen-tethered AMPs to efficiently prevent and treat bacterial wound infections.

568014 Fusogenic Liposomes Functionalized with Cell Penetrating Peptides for the Treatment of Bacterial Meningitis

Fusogenic Liposomes Functionalized with Cell-Penetrating Peptides for the Treatment of Bacterial Meningitis

Caterina Bartomeu Garcia, Di Shi, and Thomas J. Webster

Department of Chemical Engineering, Northeastern University, Boston, MA, USA

Bacterial meningitis is one of the most severe and aggressive infections that leads to permanent disabilities and brain damage to 20 percent of the population that manage to survive this disease. Bacteria spreads in the blood stream and reaches the brain site, where it provokes the inflammation of the meninges and the fluid-filled spaces in-between them. If not treated in time, bacterial meningitis can lead to permanent brain damage as well as death. The emergence of drug-resistant bacteria has increased the difficulty of treatment and the mortality rate. Bacteria with low susceptibility to current antibiotics lead to the formation of biofilms, which further compromises the ability of antibiotics to treat this disease, leading to ineffective treatment and progressed infections. The overuse of antibiotics not only promotes the proliferation of drug-resistant bacteria, but also affects other beneficial microbes in our body. To this end, there is a need to develop a more targeted drug delivery system, able to specifically interact with bacteria and combat its drug resistance.

Our research is focused on the use of fusogenic (pH-sensitive) liposomes to promote the destabilization of the bacterial cell membrane and introduce antibiotics directly inside the bacterial cell cytoplasm. Previous research reported the ability of these particles to reduce the minimum inhibitory concentration (MIC) of antibiotic necessary to kill bacteria, as well as their ability to encapsulate different type of antibiotics. For this project, our liposome systems were functionalized with a cell-penetrating peptide TAT (47-57) to increase their permeability through the blood brain barrier (BBB), one of the main obstacles that modern drugs face for treating brain diseases.

Liposomes were synthesized by the lipid film rehydration method and functionalized with TAT (47-57) using wet chemistry. Liposomes were characterized using transmission electron microscopy to verify their successful formation and size (approximately 100 nm). They were loaded with one of three commonly used antibiotics to treat meningitis, including vancomycin, methicillin, and ampicillin. These were tested against the three main bacteria that cause this disease, *Streptococcus pneumoniae*, methicillin-resistant *Staphylococcus aureus* (MRSA), and *Escherichia coli*. Growth curve assays, colony forming units, and live/dead fluorescent staining assays were performed to study the antibacterial properties of our liposomes against the previously mentioned bacteria. Liposomes were also tested against astrocytes and endothelial cells, the main cells that form the BBB, in order to demonstrate their non-cytotoxicity properties. In addition, *in vitro* BBB studies using a static model were used to test the ability of these particles to cross the BBB. These static models consist of astrocytes seeded at the bottom of a 24-well plate, with inserts containing endothelial cells added to each well. After a set incubation time, endothelial cells were placed inside a separate 24-well plate containing bacteria. Liposomes were added to study their permeability through the endothelial cell barrier to reach the bacteria.

Antibacterial results demonstrated significant antibacterial activity, especially against MRSA with a MIC of 1.7 µg/mL. Cytotoxicity results showed no interaction of the functionalized and non-functionalized liposomes with astrocytes and endothelial cells, as well as a significant reduction of antibiotic cytotoxicity through encapsulation. Initial BBB *in vitro* studies suggested a high permeability of these particles, as the concentration of bacteria across the endothelial cells was significantly lower when in the presence of liposomes.

In conclusion, our liposome system demonstrated significant antibacterial properties, especially towards MRSA and *S. pneumoniae*, while non-cytotoxicity was observed against astrocytes and endothelial cells. Furthermore, the ability of these particles to cross the BBB and combat the bacteria was successfully proved with a static BBB model, while also avoiding the destabilization of the endothelial cell membrane.

Future research is focused on the interaction of these liposomes with the bacterial cell membrane in order to better understand the bacterial cell uptake mechanism. Scanning electron microscopy (SEM) was performed to observe the bacterial cell membrane after their interaction with the liposome particles (Figure 1). This figure illustrates the disruption of the bacterial cell membrane with a low concentration of liposomes, as well as a total disruption of the bacterial cell membrane with higher concentration of antibiotics. Bacterial cell protein studies will also be performed, with a focus on the penicillin binding protein 2a, known for causing bacterial resistance to antibiotics.

This work was supported by funds from Northeastern University.

Figure 1. SEM of (A) MRSA, (B) MRSA with TAT liposomes loaded with 2 μg/mL methicillin, (C) MRSA with TAT liposomes loaded with 4 μg/mL

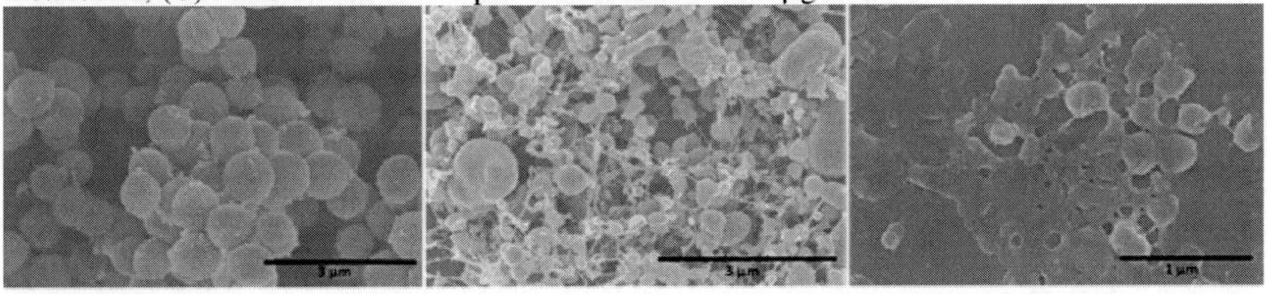

methicillin.

576224 Eradication of Bacterial Persister Cells By Targeting Membrane Potential

Bacteria are well known to enter dormancy and form persister cells which is a subpopulation exhibiting high-level tolerance to antibiotics. This feature allows persister cells to survive the treatment of potent antibiotic and cause reoccurring infections. Despite extensive research, controlling persister cells still remains a challenge. Because persister cells have a lower membrane potential than normal cells, we hypothesize that persister cells have reduced efflux activities, which can lead to higher accumulation of antibiotics intracellularly. To test this hypothesis, we compared the concentration of minocycline, a semi-synthetic tetracycline antibiotic, in normal and persister cells of *Escherichia coli* (*E. coli*) HM22. We show that with a concentration of 100μg/mL of minocycline we can kill *E.coli* persister cells by 70.8 ± 5.9% while it only killed normal cells by 10.3 ± 3.7%. In addition, the results show that persister cells accumulated ~4 times more minocycline per cell compared to normal cell. Consistently, the results were corroborated with tests using efflux pump mutants and efflux pump inhibitors. Specifically, the acrAB mutant (of the Resistance Nodulation Division efflux pumps) exhibited an increased susceptibility to minocycline; e.g. minocycline killed the acrA mutant by 99.0 ± 0.1% at a concentration of 100μg/mL. Consistently, treatment with carbonyl cyanide m-chlorophenylhydrazone (CCCP), an efflux pump inhibitor, at 10μM led to 94.7 ± 2.5% killing of *E.coli* normal cells by minocycline. This is a ~4 times increase compared to treatment without membrane depotentiation which only led to 22.0 ± 3.3% killing. Collectively, these findings reveal that persister formation may cause enhanced penetration for certain antibiotics which can lead to more killing of tolerant population under normal conditions.

565979 Mucoid Switch Enables a Growth Advantage to *Pseudomonas Aeruginosa*in Interfacial Environments

Chronic lung infection with bacterial biofilms is one of the leading causes of death in cystic fibrosis (CF) patients. Among many species colonizing the lung airways, *Pseudomonas aeruginosa* can undergo pathoadaptation leading to a mucoid phenotype with unique material properties. We hypothesize that the mucoid switch provides a growth advantage to *P. aeruginosa* through the development of interfacial films with viscoelastic properties enabling colony restructuring and cell survival.In this study, we investigated the interfacial properties of films formed by non-mucoid (PANT) and mucoid (PASL) strains of *P. aeruginosa*isolated from CF patients. We used pendant drop elastometry to analyze the interfacial response of the films formed by PANT and PASL at hexadecane-water interfaces. The dynamic rheological properties of these films exposed to interfacial stresses demonstrated that the mucoid strains have a distinctive signature. When exposed to a constant strain, films of PASL exhibit more relaxation under compression than tension, while the perfect hysteresis behavior is achieved by PANT, which is characterized by a viscoelastic response conserved under both compression and tension. Wrinkling and shape analyses of the interfacial films elucidated that mucoid strain exhibits remarkable viscoelastic properties enabling remodeling of the living films through dissipation of compressive stresses. Microscopic analysis revealed that the interfacial films of PASL retained an intact cellular morphology compared to PANT, which harbor a large number of lysed cells. This indicates that the secretion of mucoid materials by PASL plays an important role in protecting the bacteria from interfacial stresses. Further characterization of biological materials will provide new insights toward the development of methods for controlling bacterial growth on biointerfaces.

580989 Biofilm Prevention By UVC Side Emitting Optical Fibers

I have two main career goals that are related to (i) research and (ii) teaching. The first is to develop chemical free point of use water treatment device to ensure quality drinking water in both the United States and developing countries. I will work at the interface of basic science and industry to develop innovative and green technologies. The second goal is to educate diverse communities in STEM through a focus on water treatment and water resources. I will become a tenure track professor to both educate and develop a research team with a focus on photo-assisted water treatment.

Research

The lack of appropriate disinfection strategies results in an estimated 502,000 diarrheal deaths each year. Germicidal ultraviolet-C (UVC) irradiation does not produce chlorinated DBPs and requires only electrical power for disinfection. UVC Light emitting diodes (LED) have promising characteristics for water treatment (e.g. lack of warm up time, tunable radiation, no degradation at on/off cycles and longer life of use). However, their small surface area of irradiation limits their integration into treatment reactors. To progress development and integration of photon driven water treatment my research interests includes (i) better distributing light from UV lamps, (ii) improving the quantum yields of reactions, and (iii) understanding consumer perception and desires related to water treatment technologies.

First I seek to integrate optical technologies into water treatment to increase uniform light delivery. By using a UV-C glowing optical fiber (Provisional Patent), I illustrated an increase of > 600 times in surface area of irradiation as part of my dissertation. As a tenure track faculty, I will develop point of use devices that use these optical fibers, and other highly reflective surfaces in collaboration with current industrial partners (i.e. AquaSense). **Second, I will improve reaction quantum efficiency by fundamentally understanding wavelength specific photolysis mechanisms.** Proteins, DNA, and organic bonds (pharmaceuticals) are prone to lysis at specific frequencies. By using LEDs, I will explore single and synergistic wavelength effects on both microorganisms and organic contaminants of interest. This understanding will lead to development of highly efficient LED water treatment reactors. **Last, I seek to understand how community perception of water quality effects technology adaption and integration of point of use systems.** I have established strong relationships with the director of health a downriver community of Colombia (Mompox). I have and will continue to work with them to ensure the technologies I develop are truly beneficial to consumers who do not have access to reliable municipally treated water. My research will transpose into (i) advancement of water treatment science and technology and (ii) increased quality of life in developing communities.

Teaching

My main contribution as a professor will be to train the next generation of engineers to be leaders in both industry and academia. A tenure track position will enable me to accomplish this through both mentoring and teaching. By developing a research team, I will **mentor** graduate students to grow as well-rounded researchers and humans. I believe a dissertation or thesis is one piece of the graduate degree pie. I will encourage my students to fill the remaining pieces through opportunities such (but not limited to) (i) national and international internships (ii) leading outreach and service events, (iii) communicating their work with the general public, (iv) and organizing conference sessions. In addition to developing crucial soft skills, these opportunities facilitate knowledge transfer by allowing the student to better understand and communicate the broader impact of their fundamental research.

I hope to **teach** fundamental chemical engineering courses to tie technical concepts to my **application driven teaching philosophy.** I will challenge students to understand ethical and political dilemmas that are directly tied to technical classroom lectures. For example, in unit operations, I will use power plant CO2 scrubbers as examples to drive equation development of gas absorption. I will illustrate how manipulating the equation variables could influence both the efficiency and cost of CO2 scrubbers. I

will end the lecture with a brief discussion how politics has both encouraged and limited their integration in power plants over the years.

I will use the same philosophy in developing courses in nanotechnology, photo-assisted reactions, and advanced water treatment. My global experience with water treatment, strong industrial network, and active participation in the NSF Nanotechnology Enabled Water Treatment (NEWT) allow me to add impactful examples and discussions in these topics. The critical thinking skills developed through a chemical engineering degree is unparallel to any other. Those fundamental concepts rooted in broader impact can empower students to solve global problems.

565200 Invited Speaker: Understanding Virus Surface Interactions and Stability

Better understanding of virus surface chemistry could lead to improved virus removal techniques, purification methods for viral therapeutics and increased therapeutic stability. To enhance the toolbox of available characterization tools, chemical force microscopy (CFM) is being pioneered for viral particle characterization. CFM measures the adhesion force between a particle, in this case a virus particle, and a functionalized atomic force microscopy (AFM) tip. CFM reduces many of the difficulties of bulk characterization techniques by measuring adhesion of individual virus particles, likely reducing purification affects that can change surface properties. We have used CFM to study the electrostatic and hydrophobic interactions of two model viruses, porcine parvovirus (PPV) and bovine viral diarrhea virus (BVDV). These viruses were chosen because one is non-enveloped and one is enveloped, demonstrating the wide applicability of this method.

Knowledge of the electrostatic differences between the two viruses was used to thermally stabilize the virus particles. This could potentially reduce the need for a cold chain for live attenuated vaccines. The cold chain requires materials to remain below 8°C during transport. The viruses were stabilized by encapsulating them in an electrolyte dense phase. The oppositely charged polypeptides create the dense, coacervate phase. The coacervate is able to capture all detectable virus and can thermally stabilize PPV. The electrostatic differences between the viruses is noted in a shift in the charge fraction needed to encapsulate the two viruses studied.

Taken together, we have used virus surface characterization and virus polypeptide encapsulation to stabilize viruses. Future work will explore other virus surfaces to determine if knowledge of the virus surface electrostatics can be used to engineer the coacervate phase with minimal experiments.

572316 Preventing Bacterial Biofilm Formation Using Active Surface Topography

Huan Gu[1, 2], Sang Won Lee[1,2], Dacheng Ren[1,2,3,4*]

[1]Department of Biomedical and Chemical Engineering, [2]Syracuse Biomaterials Institute, [3]Department of Civil and Environmental Engineering, [4]Department of Biology, Syracuse University, Syracuse, NY 13244, United States

Bacterial adhesion on biotic or abiotic surfaces and the subsequent formation of multicellular biofilms cause persistent biofouling in industrial settings and drug-resistant infections in humans. Biofilm-associated challenges have stimulated extensive research to develop effective strategies for biofilm prevention and removal. However, most of the technologies to date are based on static features and thus are only effective for short-term control. To address this challenge, we developed a new method to repel bacteria with actively moving topographic structures, made with micron-scale poly(dimethylsiloxane) (PDMS) pillars with superparamagnetic Fe_3O_4 nanoparticles secured at the tip of each pillar. The movement of pillars was controlled by a programmable external electromagnetic field. The antifouling effects of cylindrical-shaped pillars with a diameter of 2 µm, height of 10 µm, and stiffness (Young's modulus) of 2.61 MPa, was optimized by varying the inter-pattern distance as 2, 5, 10, 15 or 20 µm. The inter-pattern distance of 5 µm exhibited the best activity in the prevention and removal of *Pseudomonas aeruginosa* PAO1 biofilms. The continuous movement of such topographic features for 6 h showed 95.3 ± 0.7 % reduction of *P. aeruginosa* PAO1 biofilm formation compared to static pillars. On-demand actuation of these pillars for 3 min detached mature (48 h) *P. aeruginosa* PAO1 biofilms by 99.9 ± 0.1%. With the capability to prevent bacterial adhesion and remove mature biofilms, this technology has potential applications in engineering smart medical devices, e.g., antifouling catheters.

564829 Invited Speaker: Developing Tools to Study and Engineer Microbes Beneficial to Human Health

Microbiota health is linked to human development and wellness, and in turn, microbial dysbiosis can stem from, or spur, disease. With the largest microbial community present in the gastrointestinal (GI) tract, a perturbed gut microbiota can invite opportunistic overgrowth, nosocomial infection, immunodysregulation, and metabolic imbalances, each of which have been associated with a multitude of diseases. The gut microbiota has been further associated with aging, asthma and atopy, cardiovascular disease, neuropsychiatric disorders, cancer, and childhood diseases. Although historically contextualized within traditional medicine and home remedies, recent explorations into treatments of these conditions have sought to leverage the microbe–microbe–host connection by restoring homeostasis through the supplementation of good bugs, colloquially termed probiotics. Lactic acid bacteria, largely from the order *Lactobacillales* and long used in food and agricultural processing, have become a centerpiece of probiotics research. While probiotic lactic acid bacteria are often sold as over-the-counter health supplements, they have an opaque connection to wellness, resulting in symptom alleviation or general health promotion without a designated mode-of-action. In this talk, I will highlight my group's recent progress in developing of tools and methodologies to understand and engineer the interactions of lactic acid bacteria with their host and other microbes with the long-term goal of demystifying the mechanisms by which probiotics promote health benefits and how to further enhance their use as living therapeutics to treat disease indications.

579656 Invited Speaker: Designing Functional Nanoculture Systems for Studying Microbial Dynamics

The diversity and abundance of species in the human microbiome have significant implications in the disease process. For instance, microbial communities in healthy mouth and gastrointestinal (GI) habitats can be affected by microenvironmental changes (e.g. temperature, humidity, pH, and chemical composition) that lead to unbalanced interactions promoting the growth of pathogens with consequences including periodontal diseases, C difficile infection, cancer, sepsis, and death. However, tools to characterize and model the signatures of healthy or diseased human microbiomes are lacking, in part due to the inability to culture some species implicated in such dynamics or define the exact cell interactions. Here, we propose to develop synthetic communities modeling human microbiome. Using a high-throughput microfluidic-based technique, we generate microbial communities in artificial microniches (nanocultures) to grow microbes with challenging culturability. Each nanoculture begins as a several nanoliter droplet of suspended cells, encapsulated by a polydimethylsiloxane (PDMS) membranes. The PDMS shell provides long-lasting mechanical support, and transport properties enabling long term study of microbial dynamics in controlled microenvironments. The resulting system is used to investigate the physicochemical interactions involved in microbial pathophysiology, microbial persistence, and antibiotic resistance.

576286 Microsphere Hydrogel Encapsulation to Identify Interactions in Chronic Wound Microbial Consortia

Most bacterial cells are members of diverse microbial communities. Community life allows for chemical and physical interactions that give rise to emergent traits such as antibiotic resistance, increased biomass production, or the expansion of tolerable growth environments. Community-associated traits play a detrimental role in chronic infections, such as cystic fibrosis lung infections, diabetic foot ulcers, and chronic rhinosinusitis, all of which are frequently caused by a polymicrobial consortia. Understanding how microbial interactions lead to observable traits in pathogenic communities is important in developing treatments for chronic infections of the human body. An especially ubiquitous and potent pathogenic consortium in chronic infections is that of *Pseudomonas aeruginosa* and *Staphylococcus aureus*. Preliminary research has identified emergent traits in this community related to pathogenicity. We aim to probe species-species interactions to further identify the emergent traits of *P. aeruginosa* and *S. aureus* consortia. These studies are conducted by encapsulating bacteria in chemically permeable hydrogel microspheres. Cells in the microspheres are physically, but not chemically, isolated. Comparing systems that allow only chemical interactions to systems that allow both physical and chemical interactions, such as bulk growth or co-encapsulation, tests the role of both interaction types. The test systems are probed metabolically and genetically. When coupled with the ability to monitor single cell growth using confocal laser scanning microscopy, and to expose cells to stresses such as oxygen limitation and antibiotic treatment, we can relate interactions to observed changes in growth and stress response. Thus, by applying a novel microsphere hydrogel system to the study of *P. aeruginosa* and *S. aureus*, we can begin to unravel the relationships between interactions and pathology in *P. aeruginosa* and *S. aureus* communities.

569847 Invited Speaker: Drop Stabilization on a Chip (DropSOAC): Stabilizing Microfluidic Drops for Time-Lapse Quantification of Single-Cell Bacterial Physiology

The physiological heterogeneity of cells within a microbial population imparts resilience to stresses such as antimicrobial treatments and nutrient limitation. Cellular heterogeneity can allow for a subpopulation of cells to survive such stresses and regenerate the community. Microfluidic approaches provide a means to study microbial physiology and bacterial heterogeneity at the single cell level, improving our ability to isolate and examine these subpopulations. Specifically, drop-based microfluidics provides a high-throughput approach to examine individual cell physiology within bacterial populations. Here, we present a simple and high-throughput method, called DropSOAC (Drop Stabilization On A Chip), that allows for imaging of individual bacterial cells within an array of microscale drops over time. The drops are less than 20 μm in diameter and remain stable as a packed monolayer under time-lapse imaging for over 20 hours. Approximately one hundred thousand individual drops can be incubated within a DropSOAC microfluidic device. Using DropSOAC, we demonstrate that growth rate and lag time heterogeneity of hundreds of individual bacterial cells can be determined starting from single isolated cells. We characterize growth of *Pseudomonas aeruginosa* and its *Δhpf* mutant derivative during resuscitation and growth following starvation. The results show that the DropSOAC method provides a high-throughput approach toward studies of microbial physiology at the single cell level, and can be used to characterize physiological differences of cells from within a larger population. This work provides a deeper understanding of drop emulsion stability within a porous PDMS microfluidic device and will have future impact in studies of the physiological heterogeneity of single cells within other microbial populations.

573176 Invited Speaker: Developing and Applying a Microdroplet Co-Cultivation and Omics Toolbox for Elucidating Microbiomes

We have been developing and applying a technological pipeline, based on nanoliter-scale microfluidic droplets, to co-cultivate and dissect subsets of complex microbial communities in order to elucidate molecular mechanisms underlying their structures and functions. The pipeline consists of droplet generation, co-cultivation, isolation, and analysis such as metagenomic sequencing. We apply this technological framework to the study of microbial communities closely related to human health, particularly the vaginal and gut microbiome. For instance, we demonstrate the anaerobic co-cultivation of diverse sub-communities from a human fecal sample and the *de novo* reconstruction of genomes from a microdroplet. We observe that many of the cultivated bacteria in these encapsulated sub-communities represent the microbial "dark matter".

581658 Invited Speaker: Efforts Toward Measurement Assurance for Microbial Biofilm Characterization

Detrimental microbial biofilms (e.g., in human infections, medical device infections, water line contamination, etc.) and beneficial microbial biofilms (e.g., for bioremediation, probiotic effects, etc.) are often considered separately, even having completely different conferences devoted to each category. However, these two broad categories of biofilms, in addition to being united by their focus on the biofilm growth state, are also united by their need for well-characterized, robust, reproducible methods to quantify biofilm properties and response to their environment.

At NIST, our biofilm efforts focus on developing the underpinning measurement science and technologies to enable quantitative characterization of biofilms. This measurement infrastructure could include tools such as protocols, best practices, uncertainty analyses, measurement methods, documentary standards, reference datasets, and reference materials. Proper usage of these tools can help decision-makers understand the level of confidence they should have in results from a measurement. In addition, these tools should also increase the overall level of confidence in biofilm measurements. Ultimately, biofilm characterization results will better inform decision-making, ranging from decisions made during research and development of new antimicrobial technologies and microbial diagnostics and therapeutics, to decisions related to regulatory oversight and medical diagnosis/treatment.

This presentation will highlight ongoing efforts at NIST focused on improved methods to characterize microbial biofilms, including cell enumeration methods and development of cell-based reference materials, fluorescence lifetime imaging coupled with machine learning to characterize cell state, and evaluation of sources of measurement uncertainty in common biofilm characterization methods.

579655 Invited Speaker: Microbial Biofilms: Structure, Transport, and Dynamics

Most micro-organisms, including bacteria, fungi, and archaea, exist in surface-adherent, multicellular communities known as "biofilms". Microbes in these films experience enhanced antibiotic resistance and improved resource processing, among other benefits. Biofilms are crucial for global carbon and nitrogen cycling and the production of industrial bioproducts and are also implicated in myriad problems including biomedical infections, food and drug contamination, and oil pipeline corrosion. Yet, despite their widespread importance, the physical properties and collective dynamics of microbial biofilmsremain poorly understood. Here, I will describe our experimental efforts to understand biofilm mechanical properties and community dynamics, as well as our recent efforts to develop technologies for structuring synthetic biofilms.

579653 Invited Speaker: Establishing and Exploiting Biocolloidal Properties of Biofilms

Bacterial biofilms are structured communities of cells encapsulated in matrix materials that include polysaccharides, proteins, and DNA. Studying bacterial biofilms as a soft matter system—where the cells are analogous to colloidal particles and the matrix materials are analogous to viscoelastic hydrogels—allows for relationships between biofilm structure and mechanics to be revealed. Structural and mechanical properties of biofilms are important to nutrient and antimicrobial transport within biofilms, biofilm fragmentation and resilience in fluid flow, and biofilm sociobiology. Here I use a biocolloidal lens to establish the microstructure of staphylococcal biofilms and to demonstrate how bacterial cells and biofilm matrix materials interact to generate biofilm morphology and mechanics. I show that staphylococcal biofilm cellular clustering and separations depend significantly on growth environment, where unstressed staphylococcal biofilms have densely packed, disordered cellular microstructures, and stressed staphylococcal biofilms contain open, tenuous microstructures. Next, artificial biofilms are created with microstructures and mechanical behaviors matching those of natural staphylococcal biofilms. These artificial biofilms are utilized to reveal the importance of pH and matrix interactions to biofilm mechanics. Natural staphylococcal biofilms are then softened through tuning the pH of their microenvironment. Findings from these studies have implications for studying biophysical contributions to staphylococcal biofilm disassembly and dispersion as well as the development of biofilm control strategies targeting matrix interactions.

574686 Inoculum Concentration As an Independent Variable to Evaluate Biofilm Structure-Function Relationships

Material selection for implanted medical devices is a challenging problem with as many as 70 % of hospital-acquired infections attributed to devices. Ideally, materials would fully integrate into the host with minimal fibrous capsule formation or immune response. However, complete integration means the material will be continuously exposed to any microbes present in host fluids. "Non-fouling" materials that use a variety of mechanisms to resist microbial attachment over the lifetime of the device are promising, but depending on material and cellular properties, the surface coverage profile of any microbes that do adhere could vary greatly from sparse to high coverage. This initial surface density is expected to impact the structure of the biofilm as it forms, which could in turn modulate the biofilm functional properties such as susceptibility to antimicrobial treatments.

Our objective was to develop a simple model system to mimic variations in cell attachment density and evaluate how the initial density relates to biofilm structure and function. *Streptococcus mutans* (*S. mutans*) was selected as a representative biofilm-forming bacterium and is a commensal oral bacterium that adheres to tooth and dental materials where it contributes to caries formation. We varied the inoculum concentration of *S. mutans* over four orders of magnitude and then quantified the resultant biofilm structures at 48 h using biofilm characterization methods based on colorimetric reporters (crystal violet, MTT (3-(4,5-Dimethylthiazol-2-yl)-2,5-Diphenyltetrazolium Bromide), coomassie blue) and fluorescent imaging (Syto9). Changes in the starting concentration of dispersed cells produced initial surface coverage profiles with distinct differences appearing as early as 6 h after inoculation, with higher inoculum concentrations leading to more surface coverage. These cell surface densities led to different biofilm structures, with higher inoculum concentrations yielding sheet-like biofilms with a homogenous appearance and lower concentrations resulting in isolated microcolonies with tower-like structures. These visible differences were maintained over a 72 h culture period. To evaluate the functional impact of the structural profiles, biofilms were subject to antibiotic challenge at various concentrations (erythromycin), and metabolic activity was measured using the MTT assay immediately after antibiotic exposure and after a 24 h recovery period with fresh medium. Based on the potential differences in antibiotic access to the *S. mutans* due to the different biofilm structures, our hypothesis was that there would be an inoculum concentration-dependent response to antibiotic challenge.

Most measures of the biofilms showed increased biofilm activity and component staining as the inoculum concentration increased. However, the antibiotic challenge did not reflect these changes and presented a more complicated picture. Inoculum concentration (and the corresponding biofilm structure) did not seem related to metabolic activity immediately after the challenge, where even 100 μg/mL erythromycin only somewhat reduced biofilm activity for all biofilms. Interestingly, providing a 24 h recovery period revealed a different response to the antibiotic challenge, with a much greater reduction in biofilm activity at the higher antibiotic concentrations and a potential dependence on inoculum concentration for some antibiotic levels. These results suggest that an immediate MTT assay may not fully represent the biofilm response to the erythromycin treatment.

The key finding from this work is that seeding concentration alone can result in large differences in initial cell attachment on a surface, which leads to differences in biofilm appearance and may affect response to antibiotics. Engineering biofilms via inoculum concentration is a promising approach to modulate biofilm structure and to develop model systems that mimic biofilms that might develop on low-attachment, "non-fouling" surfaces. Moreover, this model system could also be used for broader studies of biofilm structure-function relationships especially in the context of recovery from antibiotic treatment.

Acknowledgements: KP and HL are supported by the NIST National Research Council Postdoctoral Research Associateship Program.

565153 Invited Speaker: Biofilm Rheology

Under appropriate growth conditions, bacterial cultures will transition to forming a biofilm. To a rheologist, a scientist who studies the flow of matter, this biofilm can be viewed as a very complicated, weak hydrogel, which is composed of a complex conglomeration of bacterial clusters and extrapolymeric substances. As this structure serves to protect and insulate bacteria from mechanical disturbances and environmental perturbations, such as the intentional introduction of antibiotics, measuring the rheological properties of the biofilm at appropriate length scales can provide quantitative structural information. These characteristics can be used to evaluate the efficacy of techniques that are often directed at their removal or prevention. In this talk, we will outline the current state of the field of biofilm rheology, providing the audience with a brief introduction to the field of rheology, focusing on the benefits and shortcomings of various techniques to assess this complicated material, and discussing future directions that we need to take as a community to better understand these complex microorganism structures.

562704 Invited Speaker: Mechanical Principles of Biofilm Formation

Biofilms are surface-attached communities of bacteria that can cause problems including medical infections, fouling, and clogging in industrial applications. By contrast, beneficial biofilms are crucial in applications including waste-water treatment and microbial fuel cells. In this talk, I will discuss about our recently progress in using *Vibrio cholerae* as a model biofilm former to reveal the mechanical principles behind biofilm formation, both at the single cell level and at the continuum level. I will first present a new methodology to image living, growing bacterial biofilms at single-cell resolution, and demonstrate how cell-cell adhesion and cell-surface adhesion balance each other to cause *V. cholerae* to form an ordered, three-dimensional cluster. Next, I will show how extracellular polysaccharides, proteins, and cells function together to define biofilm mechanical and interfacial properties. Finally, I will present various mechanical instabilities that take place when biofilms grow on soft substrates, and how such instabilities, together with interfacial properties, define the morphogenesis process of bacterial biofilms.

575549 Invited Speaker: Role of Chemotactic Adaptation in *H. Pylori* Invasion and Colonization

Helicobacter pylori, a gram-negative bacterium is a leading cause of gastric cancer (non-cardia gastric cancer), and it significantly increases the risk of ulcers, gastric mucosa-associated lymphoid tissue lymphoma and gastric cancers. Typically, the lack of early symptoms delays the detection of gastric cancer till it invades muscularis propria, consistent with the poor survival rates. It is estimated that almost 60% of human beings carry these bugs, and colonization of the gastric environment typically persists throughout the life of an un-medicated host. Studies have suggested that motility and chemotaxis in *H. pylori* are key virulence factors, likely enabling cells to invade and subsequently colonize the stomach-epithelia. However, the genes that encode the enzymes responsible for chemotactic adaptation, which are crucial for temporal detection of chemical concentrations, are missing in *H. pylori*. This is at odds with recent reports that indicate that *H. pylori* undergo chemotaxis. I will discuss ongoing efforts in our lab to determine if chemotactic adaptation is possible in *H. pylori*, despite the lack of adaptation enzymes. I will present a novel approach for quantitative measurements of chemotactic activity from swimming trajectories of individual cells near surfaces. Through a combination of optical trapping that enables single motor assays, phase microscopy, and dark field-based sensitive detection of flagellar rotations, we have observed anisotropic rotation modes in flagellar bundles. We have exploited these modes to overcome current technical limitations that prevent single cell measurements. These modes appear to be arise consistently across bacterial species. We anticipate that our results will facilitate the development of antivirulence strategies to minimize the potential for invasion and colonization by *H. pylori*.

574738 Diatom Microbubbler for Active Biofilm Removal in Confined Spaces

Bacterial biofilms are formed on many living tissues, engineered materials, and medical devices, which threatens human health and sustainability. Although tremendous efforts have been made on biofilm removal, it is still a grand challenge to remove biofilms formed in confined space. One primary cause is extracellular polymeric substances (EPS) of the biofilm, which limit the transport of antibacterial agents into biofilm. In this work, we hypothesized that microparticles engineered to produce microbubbles with self-locomotion would remove biofilms by fracturing the EPS and subsequently enhancing transports of the antiseptic reagent. We examined this hypothesis by doping a porous cylinder-shaped diatom biosilica particle with manganese oxide (MnO_2) nanosheets. In antiseptic H_2O_2 solution, the diatoms doped with MnO_2 nanosheets, denoted as diatom microbubbler, continuously discharged oxygen gas microbubbles and became self-motile. Subsequently, the diatoms penetrated the bacterial biofilm formed on microgrooved poly(dimethylsiloxane) (PDMS) substrates and continued to generate microbubbles. The resulting microbubbles merged and converted surface energy to mechanical energy to fracture the EPS matrix in biofilm. Consequently, H_2O_2 molecules diffused into the biofilm and killed most bacterial cells. Overall, this study provides a powerful and unique tool that can significantly impact current efforts to clean a wide array of biofouled products and devices.

580140 Invited Speaker: The Bacterial Sense of Touch: Chemical and Mechanical Interactions of Bacterial Cells at Interfaces

An important feature of biofilms is the phenotypic difference of cells within the community from those free in solution (planktonic cells). Part of these differences arise from chemical communications (quorum sensing) between closely situated cells on a surface and within the film. However, evidence suggests that even before cells accumulate with sufficient density for a quorum sensing mechanism, bacteria sense the interfacial environment (different from that in solution) through other means. This talk explores the mechanical deformations of interfacial bacterial cells that occur as a result of their adhesion to model surfaces. We share early data on the effects of these interactions on the cells themselves: in some cases cell death is accelerated but in other cases, cells grow. These findings lay the groundwork for material design to manipulate bacterial cells, ultimately killing them or exploiting them for biological and technological benefits.

573537 Invited Speaker: How Bacteria Read the Map: Effects of Surface Topography on Biofilm Formation

Bacterial adhesion on implanted biomaterials and formation of multidrug resistant biofilms can have catastrophic impacts on the infected individuals. As a major material property, surface topography plays an important role in biofilm formation. In this presentation, we will prevent a brief overview of this topic followed by a few of our studies aiming to control biofouling by engineering surface topography. By investigating bacterial adhesion on polymers with specific topographic patterns, we proposed a set of principles for rational design of antifouling surfaces and validated the design using protruding hexagonal patterns, which were found reduce biofilm formation of *Escherichia coli* by around 90%. Inspired by these findings, we further developed a new strategy to remove established biofilms using biocompatible shape memory polymers with micron-scale topographies. We demonstrate that this strategy can achieve a total reduction of *Pseudomonas aeruginosa* biofilms by 99.9% compared to the static flat control. It was also found effective against biofilms of *Staphylococcus aureus* and an uropathogenic strain of *Escherichia coli*. The detached cells were found to be more sensitive to antibiotics than the original biofilm cells, demonstrating synergic effects in biofilm control. These results inspired further improvement to develop biofilm control strategies with reversible shape memory polymers and active topography with beating pillars driven by electromagnetic actuation. Active topography was found to both inhibit biofilm formation and remove mature biofilms of *P. aeruginosa* by 99.9%, offering more prolonged control of biofouling. These results can help design better medical devices to prevent biofilm infections.

574471 Invited Speaker: Motility and Adhesion of Bacteria Near Oil Droplets

The interactions of bacteria with the oil/water interface of dispersed droplets plays an important role in numerous technological processes, including bioremediation of oil spills as well as droplet microfluidics for pathogen detection, antibiotic susceptibility, and biotechnological selection. Here, I will describe our recent studies in which we combine confocal microscopy and microfluidic emulsification to investigate interactions of bacteria with oil droplets suspended in water and stabilized by surfactants. We employ these techniques to probe the adhesion of both non-motile and motile bacteria on the droplet surface over time, examining the role of surfactant type and concentration as well as bacterial motility. Briefly, we find that adhesion is controlled by the interfacial tension, and the ability to swim enhances adhesion at the droplet interface. Interactions with a boundary can generate more complex behaviors. In this context, we show that motile bacteria can rotate oil droplets resting near a solid surface. This rotation is correlated both to cell adhesion to oil droplets and to cell motility. The methods applied here can be used to optimize and tune the interactions of other bacteria with interfaces in a variety of technologically relevant settings.

572203 Invited Speaker: Transport Phenomena of Microbes in Groundwater and Porous Media

Nonaqueous phase liquid (NAPL) contaminants are difficult to eliminate from natural aquifers due, in part, to the heterogeneous structure of the soil matrix. Residual NAPL ganglia remain trapped in regions where the hydraulic conductivity is relatively low. Bioremediation processes depend on adequate mixing of microbial populations and the groundwater contaminants that they degrade. The ability of microbes to sense a chemical gradient and swim preferentially toward locations of higher concentration, known as chemotaxis, can enhance the mixing of microbes with contaminant sources that may not be readily accessible by advection and dispersion alone. Accumulation of microbes at the oil-water interface enhances the rate at which contaminants are degraded.

Apparent diffusion and dispersion coefficients determined from the experimental observations are used in mathematical models to predict macroscopic-scale transport of bacterial populations. A dimensionless chemotaxis number is proposed to ascertain *a priori* the conditions under which a chemotactic response will impact bacterial transport relative to other processes such as advection and dispersion. This presentation will focus on observations of microbial chemotaxis in a microfluidic device and sand-packed column designed to mimic features associated with residual NAPL contamination in groundwater.

561565 Invited Speaker: Goodbye Hospitals: Hello Implantable Nano Sensors

There is an acute shortage of organs due to disease, trauma, congenital defect, and most importantly, age related maladies. While tissue engineering (and nanotechnology) has made great strides towards improving tissue growth, infection control has been largely forgotten. Critically, as a consequence, the Centers for Disease Control have predicted more deaths from antibiotic-resistant bacteria than all cancers combined by 2050. Moreover, there has been a lack of translation to real commercial products. This talk will summarize how nanotechnology can be used to increase tissue growth and decrease implant infection without using antibiotics but using sensors (while getting regulatory approval). Our group has shown that nanofeatures, nano-modifications, nanoparticles, and most importantly, nanosensors can reduce bacterial growth without using antibiotics. This talk will summarize techniques and efforts to create nanosensors for a wide range of medical and tissue engineering applications, particularly those that have received FDA approval and are currently being implanted in humans.

581347 Chemical Engineering Innovations for a Renewable Economy

Chemical engineering evolved in the twentieth century with rise in the use of fossil resources and myriad of products derived from then. An array of these products required great innovations and made huge impact on the human race, and such examples include urea, transportation fuels, polymers, water purification chemicals, etc. Going forward as we make transition from fossil resources to renewable resources such as solar energy, wind and sustainably available biomass, we again have great opportunities for innovation that would not only make great impact but redefine chemical engineering. My talk will focus on my own experiences with challenges and opportunities presented to us as chemical engineers. I will share an overall perspective on the use of biomass as sustainable source of carbon, harvesting solar energy in conjunction with growing food without any land constraint, and creating solution processed solar cells that utilize chemical engineering principles along with knowledge from other engineering disciplines.

581350 Engineering Materials for Clinical Application

Dr. Schmidt's research has resulted in numerous patents that are licensed to companies, including AxoGen, Alafair Biosciences, Smith and Nephew, and Siluria Technologies. Comments for this panel will focus on peripheral nerve applications and one key success that is the foundation for the Avance Nerve Graft from AxoGen. In addition, along the way to uncovering strategies for nerve regeneration, Dr. Schmidt and her team also discovered a biomaterials processing approach that led to a separate application of protecting tendon's after surgery, resulting in the VersaWrap Tendon Protector from Alafair Biosciences, a start-up company in Austin, Texas.

Nerve Repair: Dr. Schmidt's technology for processing decellularized nerve tissue was licensed to AxoGen in Alachua, Florida (http://www.axogeninc.com) for the AVANCE™ nerve graft. Many tens of thousands (~100,000) of AxoGen's AVANCE™ grafts have been implanted in patients at hundreds of hospitals globally.

Post-surgical Wound Management: Using a novel in situ crystallization processing technique, Dr. Schmidt's team created mechanically robust hyaluronic acid-based hydrogel films that can be easily handled, sutured, and delivered laparoscopically. This research is the basis of Alafair Biosciences, a start-up company in Austin, TX (http://www.alafairbiosciences.com), which is run by Dr. Schmidt's PhD graduate and a neurosurgeon collaborator. Alafair's first product, VersaWrap™ Tendon Protector, has been implanted in more than 500 patients since December 2017.

AUTHOR INDEX

Abbasi, Reha 17
Agrawal, Rakesh 48
Aljaafari, Haydar 16
Bahar, Ali A. 24
Ballance, William 13
Balmuri, Sricharani 25
Beckwith, Joanne 15
Brown, Angela C. 8
Bunker, Mallory 28
Camesano, Terri A. 21
Carlson, Taylor 3
Carrier, Rebecca L. 3
Chang, Connie B. 32, 33
Chang, En H. 8
Chan-Park, Mary B. 14
Chu, Sang-Hyon 42
Chu, Weiping 11
Conrad, Jacinta C. 45
Da Silva, Sandra M. 35
Davis, Christian 6
De La Fuente-Nunez, Cesar 10
Deng, Yu-Heng 42
Dunkers, Joy 35, 38
Dushnov, Vitaliy 42
Edson, Julius 11
Elhebeary, Mohamed 13
Ford, Roseanne M. 18, 46
Garcia, Caterina B. 22
Garcia-Segura, Sergi 26
Gu, Huan 24, 29
Heldt, Caryn L. 28
Hong, Seok H. 12
Hong, Yu-Tong 42
Horn, Norman 6
Howell, Caitlin 4
Hu, Yuhang 13
Joshi, Pratik U. 28
Karlsson, Amy J. 9
Kim, Dong H. 42
Kong, Hyunjoon 13, 42
Koufos, Evan 8
Kratofil, Rachel M. 1
Krueger, Eric 8
Kubes, Paul 1
Kwon, Young J. 11
Lai, Yang 13
Lanzarini-Lopes, Mariana 26
Lee, Sang 29
Lele, Pushkar 41
Leonard, Heidi 38
Leong, Jiayu 42
Li, Jianghua 14
Lin, Nancy J. 35, 38
Lin, Xiaoxia N. 34
Lock, Jaclyn 3
Lozeau, Lindsay D. 21
Ma, Tianhui 2
McClelland, Michael 11
McTigue, Whitney C. B. 28
Mi, Xue 28
Nair, Nikhil U. 30
Niepa, Tagbo H. R 25, 31
Nuxoll, Eric 7, 16
Oh, Inkyu 13
Pallister, Kyler B. 1
Park, Cheol 42
Park, Jun D. 42
Parratt, Kirsten 35, 38
Perry, Sarah L. 28
Pettygrove, Brian A. 1
Porwollik, Steffen 11
Pratt, Shawna 32
Ren, Dacheng 24, 29, 44
Rogers, Simon 42
Rolle, Marsha W. 21
Roy, Sweta 24
Russell, Petria 32
Saif, Taher 13
Saleh, Bahram 19
Santore, Maria M. 43
Schmidt, Christine 49
Seo, Yongbeom 42
Shi, Di 22
Shukla, Anita 5
Soh, Joonghui 42
Solomon, Michael J. 2, 15
Stewart, Elizabeth J. 37
Stewart, Philip S. 1
Usman, Huda 31
Van Epps, J. S. 2, 15
Voyich, Jovanka M. 1
Walker, Travis W. 39
Wang, Chia-Ming 3
Waters, Nicholas 25
Webster, Thomas 19, 22, 47
Wei, Ziqi 21
Westerhoff, Paul 26
Wilking, James 17, 36
Yan, Jing 40
Yang, Yi Y. 42
Zhao, Xueying 18